CHARTERS

OF THE

WEEHAWKEN FERRY COMPANY,

AND THE

Weehawken & Ramapo Plank Road Company,

WITH PROSPECTUS.

Subscription Books of both Companies to be opened the 9th of May, 1853, at the Otto Cottage, Hoboken.

New-York:
FRANCIS HART, PRINTER AND STATIONER.
167 WASHINGTON STREET.

1853.

CHARTERS

OF THE

WEEHAWKEN FERRY COMPANY,

AND THE

Weehawken & Ramapo Plank Road Company,

WITH PROSPECTUS.

Subscription Books of both Companies to be opened the 9th of May, 1853, at the Otto Cottage, Hoboken.

New-York:
FRANCIS HART, PRINTER AND STATIONER,
167 WASHINGTON STREET.

1853.

☞ The Commissioners to receive Subscriptions to the Capital Stock of the Ramapo and Weehawken Plank Road Company, will open books of Subscription at the Hotel of Charles Perry, known as the Otto Cottage, in Hoboken, at 10 o'clock in the forenoon, on the 9th of May, 1853.

The said Books of Subscription will remain open at said place for three days. The shares to be fifty dollars each, and five dollars on each share to be paid at the time of subscribing.

☞ Books of Subscription for the Stock of the Weehawken Ferry Company will also be opened by the proper commissioners, at the same time and place, and will remain open for three days.

AN ACT TO INCORPORATE THE WEEHAWKEN FERRY COMPANY.

---

1. Be it enacted *by the Senate and General Assembly of the State of New Jersey*, That William Cooper, Rodman M. Price, David Allerton, Francis Price, Elijah Ward, Dudley S. Gregory, Barney Bertram, David Beldam, Lorenzo W. Elder, their present and future associates, their successors and assigns, be, and they are hereby created a body corporate and politic, by the name of "The Weehawken Ferry Company," for the purpose of establishing a ferry between some suitable point or points in the city of New York and a point north of Deas's Point, in the county of Hudson, with power to build boats, wharves, piers, bulkheads, and all other improvements necessary to carry out the objects of this corporation, with the privilege of asking any reasonable rates of toll, as by the by-laws of said company may be established, such tolls not to exceed the rates now taken at the Hoboken ferry; and also, to lease, purchase, and hold such real and personal estate as may be necessary for the purposes of said corporation, and sell, lease, allot, and parcel the same, or any part thereof, in such manner as the said corporation may determine, together with all the powers and privileges, and subject to such restrictions, limitations, and conditions, as are specified in the act entitled, "An act concerning corporations," approved April fourteenth, eighteenth hundred and forty-six.

2. *And be it enacted*, That the capital stock of this company shall be five hundred thousand dollars, divided into shares of fifty dollars each; and the said shares shall be deemed and considered personal estate; and it shall be lawful for said company to grant certificates of shares, in full or part payment, for the real or personal estate that may be purchased or leased, and, by the by-laws, to compel payments of instalments, not exceeding five dollars on each share at any one time, not deemed or declared full stock; and on failure to pay any instalment, to forfeit the stock, and all previous payments made thereon, giving, at least, sixty days' notice of such call and demand in a newspaper published in the county of Hudson, and one in the city of New York.

3. *And be it enacted*, That it may be lawful for said corporation to establish a ferry from a point north of Deas's Point, in the county of Hudson, to the city of New York; to construct piers, wharves, bulkheads, or such other improvements as may be made thereon, by said corporation, and keep up and maintain the same, upon the land now covered by the water in front of the lands of said company; *provided*, the same shall not obstruct the navigation of the Hudson river, and to erect ferry houses and other buildings.

4. *And be it enacted*, That William Cooper, Rodman M. Price, David Allerton, Francis Price, Elijah Ward, Dudley S. Gregory, Barney Bertram, David Beldam, and Lorenzo W. Elder, shall be the first directors to organize and manage the affairs of said company, and

shall continue in office until the first Tuesday in October, eighteen hundred and fifty-two, and until others are elected or appointed in their stead; and the above named persons are hereby authorized to open books of subscription to the capital stock of said company, at such time and place as they may think proper, by giving, at least, two weeks' notice thereof in two newspapers published in the county of Hudson, in the State of New Jersey, or in one newspaper in said county, and one newspaper published in the city of New York; and when one hundred thousand dollars shall have been subscribed, it shall be lawful for said company to commence their operations, as authorized by this act.

5. *And be it enacted,* That the property and affairs of this company shall be managed and conducted by nine directors, being shareholders, a majority of whom, together with the secretary, shall be residents of this state; and the secretary shall keep an office in the county of Hudson; the president shall be appointed from the directors; and the directors shall have power to make all needful by-laws, not inconsistent with the laws of this state or of the United States.

6. *And be it enacted,* That the annual election of directors shall take place on the first Tuesday in October, eighteen hundred and fifty-two, at some convenient place in the county of Hudson, between the hours of twelve o'clock at noon, and three o'clock in the afternoon of that day; all elections shall be by ballot, and each share entitled to one vote; and the vote may be by person or proxy; two weeks' previous notice shall be given in a newspaper published in the county of Hudson; and if, from any cause, an election for directors shall not take place at the appointed time, it shall not therefore work a forfeiture of the charter, but a new election shall be ordered, in conformity to the by-laws of said corporation.

7. *And be it enacted,* That no transfer of stock of said corporation shall be valid or effectual until such transfer shall be entered or registered in the book or books to be kept by the president or directors for that purpose; which said book or books shall be closed for the purposes of transfer of stock fifteen days before each election for directors; and no person shall be allowed to vote upon any stock, unless the same has been transferred to him or her and registered in the said transfer books, more than fifteen days prior to the election at which he or she claim to vote; and all the books of said corporation shall be open to the inspection of the stockholders, except the said transfer books.

8. *And be it enacted,* That it shall not be lawful, at any time hereafter, for any owner, captain, or other person having control of any steamboat, or other boat or vessel, to touch or to land at said docks, wharves, piers, or bulkheads of said corporation, or to receive or to land or discharge any passengers or freight at said docks, wharves, piers, or bulkheads, or such other improvements as may be made thereon by said corporation, unless in case where any boat or other vessel shall be in distress, without consent first had and obtained from said corporation.

9. *And be it enacted,* That this act shall continue in force thirty years; and that unless said company, within five years from the passage of this act, shall have established a ferry, and have the same in operation, so as to accommodate the inhabitants, this act shall be void.

AN ACT

TO INCORPORATE THE RAMAPO AND WEEHAWKEN PLANK ROAD COMPANY, IN THE STATE OF NEW-JERSEY.

---

1. Be it enacted *by the Senate and General Assembly of the State of New Jersey*, That Dudley S. Gregory, Francis Price, Hugh Maxwell, jun., Ashbel Green, Edward Suffern, Henry M. Alexander, Rodman M. Price, Garret G. Ackerson, John Lovett, jun., and Abraham Van Horn, and such other persons as may hereafter be associated with them, shall be, and are hereby constituted a body politic and corporate in law, by the name of "The Ramapo and Weehawken Plank Road Company."

2. *And be it enacted*, That the capital stock of the said corporation shall be three hundred thousand dollars, and shall be divided into shares of fifty dollars each, which shall be deemed personal estate, and shall be transferable in such manner as the by-laws of said corporation shall direct.

3. *And be it enacted*, That the above-named persons, or a majority of them, shall open books to receive subscriptions to the capital stock of the said corporation, at such time or times, and place or places, in the county of Hudson, as they, or a majority of them, may think proper, giving notice thereof, at least twenty days prior to the opening of said books, in one or more newspapers printed in the county of Hudson; and at the times and places so fixed, the said commissioners, or a majority of them, shall attend and receive subscriptions to the capital stock; and at the time of subscribing for said stock, five dollars on each share subscribed shall be paid to the said commissioners; and the residue may be called in, and shall be paid at such times and in such amounts, by instalments, as is by this act hereinafter directed.

4. *And be it enacted*, That whenever there shall be five hundred shares of the said stock subscribed, and the amount of five dollars on each share paid in, as before directed, the said commissioners, or a majority of them, shall give notice, as above specified, for a meeting of the stockholders for the purpose of choosing directors, and organizing said company, and of which said election the said commissioners, or a majority of them, shall be the judges; which subscribers, when so met, shall proceed to elect, by ballot, from among the stockholders, nine directors, a majority of whom shall be citizens and residents of this state, and hold their office for one year, and until others are elected; and each stockholder, at such election, and at all future elections of said corporation, shall have one vote for each share they may hold at the time of such election; and that such stockholders may vote at any election by proxy duly authorized for that purpose.

5. *And be it enacted*, That a majority of the board of directors shall at all times be a quorum for the transaction of business, and may have

power to call in the remainder of the capital stock of said corporation, by instalments not exceeding fifteen dollars on each share, by giving thirty days' notice of such required instalment in one or more of the newspapers printed in the county of Hudson; and if any stockholder shall neglect or refuse, for thirty days after such instalment is due, to pay the same, he, she, or they, so neglecting or refusing, shall forfeit their stock, and all payments thereon; and the said board of directors shall and may require from the treasurer such security as to them shall seem just.

6. *And be it enacted,* That when the said board of directors are so elected and chosen, the said commissioners are hereby authorized and required to pay over to the said board of directors, or to such persons as they, or a majority of them, shall direct, all money which they have received for the subscription to said capital stock, first deducting therefrom all expenses to which they have been exposed, and the sum of one dollar and fifty cents per day for each and every day they have been employed in the duties of their appointment; and the directors so chosen, and their successors, shall annually thereafter cause an election to be held, at such time and place as their by-laws shall direct, for directors of the said corporation.

7. *And be it enacted,* That in case it should happen that an election of directors should not be made on the day or at the time when pursuant to this act it ought to be made, the said corporation shall not for that cause be deemed to be dissolved, but such election may be held at any time; and the directors for the time being shall continue to hold their offices until others shall have been chosen in their places.

8. *And be it enacted,* That the president and directors of said company are hereby authorized and invested with all the rights and powers necessary and expedient to survey, lay out, and construct a plank road, not exceeding sixty feet in width, to commence at some point at or near Ramapo, in the county of Bergen, to the ferry established by the Weehawken Ferry Company, in the township of North Bergen, in the county of Hudson, with a branch road from said road at said ferry, extending northerly to Fort Lee, in said county of Bergen; and also one other branch, extending northerly from the aforesaid main plank road, on the east side of English creek, through the English Neighborhood, not exceeding ten miles in length, on such a course as may be deemed advisable, with right to erect such bridges as may be necessary; and where bridges are built over navigable streams, with such draws on the same as are made on other bridges over such streams; and it shall and may be lawful for the said president and directors, their agents and others in their employ, to enter, at all times, upon all lands or water, for the purpose of exploring, surveying, leveling, or laying out the route or routes of such road, and of locating the same, doing no unnecessary injury to private property; and when the route of such road shall have been agreed upon, and filed in the office of the secretary of this state, then it shall be lawful for the said company, by its officers, agents, contractors, and other persons in their employ, to enter upon, take possession of, hold, have, use, occupy, and excavate any such lands, and to erect embankments, bridges, and all other works necessary to construct said road, and to do all other things which shall be suitable or necessary to carry into effect the object of this charter; and it shall

be lawful for the said company, at any time, to drive piles, and erect or build piers, wharves, platforms, ferry stairs, toll or other houses, or other works necessary for the said road, and ferries thereon, over any navigable stream crossed thereby, and also to lease, purchase, and hold such real and personal estate as may be deemed necessary for the purposes of said corporation, and sell, lease, allot, and parcel, or otherwise improve the same, or any part thereof, in such manner as the said corporation may determine; *provided always*, that the said corporation shall pay, or make tender of payment, for all damages for the occupancy of lands through which the said road shall be laid out, before the said company, or any person in their employ, shall enter upon or break ground in the premises, except for the purpose of surveying said route, unless the consent in writing of the owner or owners of said lands be first had and obtained; and the said company may contract with any turnpike or bridge company, on the route they shall locate for the said road, for the purchase or hire of such turnpike or turnpikes or bridges, or any part thereof, and the same may be leased or conveyed to them by the said turnpike or bridge company.

9. *And be it enacted*, That the roadway for travel, hereby authorized to be constructed, shall be formed of good substantial plank or boards, not less than eight feet in width, laid down in a firm and workmanlike manner, to present a smooth surface; and the same shall not rise above five degrees from the horizontal at any point.

10. *And be it enacted*, That the bridges across the Hackensack river shall be so constructed as that one part thereof shall not be less than thirty feet over, to draw for the free passage of such vessels with standing masts as shall from time to time have occasion to pass up and down the said river; and the said draws shall be so constructed, for the safe passage of vessels with standing masts, with piles, piers, and platforms on each side of the draws of the said bridges, as are in the bridges which are already built over the Hackensack river, for the free navigation of vessels through the same; and being so constructed, shall, for the term hereby granted, be supported and maintained by the said directors and their successors; and for the safety of the navigators, a lamp, sufficient for the purpose, shall be placed, at the expense of the directors and their successors, on one side of the draws, in each of said bridges, on a post for that purpose; which said lamp shall be lighted every evening thereafter, as long as the said bridge shall stand, before it grows dark, and continue lighted until daylight the next morning; and the said directors and their successors shall, as long as the bridges are standing, keep, or cause to be kept, at each bridge, a careful person, to open the draws of the said bridge for the passage of vessels with standing masts; and for each and every night's neglect for lighting the lamp, and for every neglect in opening the draw, the said directors, their agents and servants, shall pay the sum of ten dollars, to be recovered by action of debt, before either of the justices of the peace of the county of Bergen or Hudson, by any person suing for the same, one half to and for the use of the prosecutor, and the other half for the use of the poor of the township of Hackensack, in the county of Bergen, and of the township of North Bergen, in the county of Hudson.

11. *And be it enacted*, That when the company, or its agents, cannot agree with the owner or owners of such required lands or materials, for

the use or purchase thereof, or when, by reason of the legal incapacity or absence of such owner or owners, no such agreement can be made, a particular description of the lands or materials, so required for the use of the said company, in the construction or repairs of said road, shall be given in writing, under oath or affirmation of some engineer or proper agent of the company, and also the name or names of the occupant or occupants, if any there be, and of the owner or owners, if known, and their residence, if the same can be ascertained, to one of the justices of the supreme court of this state, who shall cause the said company to give notice thereof to the persons interested, if known and resident in this state, or, if unknown or out of this state, to make publication thereof, as he shall direct, for any term not less than twenty days, and to assign a particular time and place for the appointment of commissioners, hereinafter named; at which time and place, upon satisfactory evidence to him of the service or publication of such notice as aforesaid, he shall appoint, under his hand and seal, three judicious, impartial, and disinterested persons, not resident in the county in which the lands or materials in controversy lie, or the owners reside, commissioners, to examine and appraise the said lands or materials, and to assess the damages; the said commissioners are also directed and required to assess the damages which any individual may sustain by said road, arising from the removing, making, and maintaining the fencing on the line of the said route, through any improved lands through which the same may run, upon such notice, to be given to the persons interested, as shall be directed by the justice making such appointment, to be expressed therein, not less than twenty days; and it shall be the duty of the said commissioners, having first taken and subscribed an oath or affirmation before some person duly authorized to administer an oath, faithfully and impartially to examine the matter in question, and to make a true report, according to the best of their skill and understanding, to meet at the time and place appointed, and proceed to examine the said lands or materials, and make a just and equitable estimate or appraisement of the value of the said lands or materials, and assessment of the damages sustained by the owner or owners thereof, by reason of the taking the same for the use of the company, which shall be paid by the company for such lands or materials, or damages aforesaid; and the said commissioners shall make a report in writing, under their hands, or the hands of any two of them, of the value of said lands, materials, and damages, which report shall, within ten days thereafter, be filed, together with the aforesaid description of the land or materials, and the appointment and oaths and affirmation aforesaid, in the clerk's office of the county in which the said lands and materials are situate, to remain of record therein; and the said report, or a copy thereof, certified by the clerk of the said county, shall at all times be considered as plenary evidence of the right of the said company to have, hold, use, occupy, possess and enjoy the said lands or materials, and of the right of the said owner or owners to recover the amount of the said valuation and assessment, with interest and costs, in an action of debt, in any court of competent jurisdiction, in a suit to be instituted against the company, if they shall neglect or refuse to pay the same for twenty days after demand made of their treasurer, and until the same be paid, shall constitute a lien upon the property of the company in the nature of a mortgage; and the said justice of the supreme court shall, on application of either party, and on reasonable notice to the others, tax and allow such costs, fees, and expenses to the justices of the supreme court, commissioners, clerks, and other persons performing any of the duties prescribed in this section, as he or they shall think equitable and right, which costs shall be paid by the said company.

12. *And be it enacted*, That in case the said company, or the owner or owners of the said lands or materials, shall be dissatisfied with the report of the commissioners named in the preceding section, and shall apply to the justices of the supreme court, at the next term after the filing of the said report, the said court shall have power, on good cause shown, to set the same aside, and thereupon to direct a proper issue for the trial of the said matter in controversy to be formed between the said parties, and to order a jury to be struck, and a view of the premises or materials in question to be had, and the said issue to be tried at the next circuit court to be holden in the county where said lands or materials, or any part thereof, may be, in the same manner as other issues in fact are tried in the said court, upon twenty days' notice of trial and six days' notice of the view being given by either party to the other; and, upon such trial, it shall be the duty of the said jury to assess the value of the said lands or materials, and damages sustained by reason of the taking thereof, as aforesaid; and if they shall find a greater sum than the said commissioners have awarded in favor of the said owner or owners, then judgment thereon, with costs, shall be entered against the said company, and execution be awarded therefor; but if the said jury shall be applied for by the said owner or owners, and shall find the same, or a less sum than the said commissioners shall have awarded, then costs shall be paid by the said applicant or applicants, and deducted out of the said sum so found by the jury, or execution awarded therefor, as the court shall direct; *provided*, that such application for an issue shall not prevent the said company from taking and using the said lands and materials, upon the filing of the said report, and tender or payment of the sum awarded by the commissioners.

13. *And be it enacted*, That in case any owner or owners of such land or real estate shall be feme covert, under age, non compos, out of the state, or under any other legal disability which would prevent their agreement with the said company, then it shall be the duty of the said corporation to pay the amount of any award or report, so made in behalf of any such persons, or the amount found by a jury, into the court of chancery, or to the clerk thereof, subject to the order of the said court, for the use of said owners; all which proceedings, as well under this, as the last section of this act, shall be at the proper costs and charges of the said corporation, except in cases of appeal above provided for; and the said justice shall and may direct the amount of cost and charges of such valuation, appraisement, and witness' fees.

14. *And be it enacted*, That the said company may erect gates and turnpikes across the said road, whenever three miles of the said road is completed, and demand and receive toll for each mile of the said road so made, not exceeding the following rates, to wit:

For every carriage, sleigh, or sled, drawn by one beast, two cents.
For every additional beast, one cent.
For every horse and rider, or led horse or mule, one cent.

For every dozen of calves, sheep, or hogs, and so on in proportion for any greater or less number, six cents.

For every dozen of horses, mules, or cattle, and so on in proportion for any greater or less number, seven cents.

And it shall be lawful for any toll gatherer to stop any horse, mule, calves, sheep, hogs, carriage of burthen or pleasure, from passing through any of the said gates or turnpikes, until the toll, as above specified, has been paid for them respectively.

15. *And be it enacted,* That before the said company shall demand or receive toll for travelling said road, they shall cause mile stones or posts to be erected, one for each and every mile on said road; and on each stone or post shall be fairly and legibly marked the distance the said stone or post is from the point of the commencement of the said road at the said ferry; and shall cause to be affixed, and always kept up at the gates aforesaid, in some conspicuous place, a list of the rates of toll which may be lawfully demanded.

16. *And be it enacted,* That if any person shall wilfully break, throw down, or deface any of the mile stones or posts so erected on the said road, or shall wilfully cut, break down, destroy, or deface, or otherwise injure any gate, turnpike, bridge, machinery, timber, or plank, that shall be erected, built, placed, or laid down in pursuance of this act, or shall forcibly pass the gates or turnpikes without having paid the legal toll, such person or persons shall forfeit and pay a fine of twenty dollars, besides being subject to an action of damages for the same, to be recovered by the said company, to their use, in action of debt, with costs of suit; and if any person shall, with team, carriage, or horse, turn out of said road to pass a gate or gates or turnpikes, and again enter on said road with intent to avoid the toll due by virtue of this act, such person or persons shall pay three times as much as the legal toll would have been for passing through said gates, to be recovered by said company, to their use, in an action of debt, with costs of suit.

17. *And be it enacted,* That if any toll gatherer shall unnecessarily delay or hinder any person passing at any of the gates, or shall receive more toll than is by this act established, he shall for every such offence forfeit and pay the sum of twenty dollars, with costs of suit, to be prosecuted for and recovered for the sole use of the person so unreasonably hindered or defrauded.

18. *And be it enacted,* That all drivers of carriages, sleighs, or sleds, whether of burthen or pleasure, or persons on horseback, using the said road, shall keep their horses, carriages, sleighs, or sleds on the right hand of said road, in the passing directions, leaving the other side of the road free and clear for other carriages or persons on horseback to pass; and if any person shall offend against this provision, such person or persons shall, besides being liable to make compensation for all damages, forfeit and pay the sum of five dollars, to any person or persons who shall be obstructed in his or her passage, and will sue for the same, to be recovered in an action of debt, with costs of suit.

19. *And be it enacted,* That if the said plank road shall not be commenced within two years, and three miles thereof shall not be in use within five years from the passage of this act, that then and in that case this act shall be void.

Approved, March 11, 1853.

PROSPECTUS

OF THE

WEEHAWKEN FERRY COMPANY,

AND

WEEHAWKEN & RAMAPO PLANK ROAD.

The charters of *The Weehawken Ferry Company*, and *The Weehawken Plank Road Company*, have been secured from the legislature of the State of New Jersey. The western shore of the North River is rapidly peopling with the overflowings of New York, and is advancing briskly in importance. When made more accessible to denizens of the city, its progress will be accelerated. The winter season finds this shore less encumbered with ice than the eastern shore. The summer season sees it surrounded by many advantages. Elevated, fresh and pleasant, the grounds of Weehawken present an invaluable outlet to the thronging thousands of the metropolis. Seekers after salubrious air and recreation, resort hither in warm weather. The region affords many inducements as a desirable and healthful locality for residence. The state of New Jersey is not only free from debt, but receives a large annual revenue from its canals and railroads. Its public schools are flourishing, and derive abundant support from state funds in addition to the amount annually received from Township tax. In Hudson county, several of the schools are already free, and others are upon the verge of this condition.

Taxes in this county are much less than in New York, and this reason, with others, is leading many to escape from the din and dust of the city to more congenial homes on the other side of the river. The position of the Ferry Depot on the New Jersey side will be opposite a point between 50th and 60th streets, New York City. Ample support, particularly in the pleasant season, will be early afforded to it. Agricultural and manufacturing products in great quantities, will be annually conveyed over this ferry, as well as many thousands of cattle destined to the city market. The westerly portions of New York City have of late increased with marked rapidity; they show great thrift and growth in opulence. The Hudson flows to its doors with the main exports of the rich regions through which it flows. From these wharves embark the major portion of our steamers. In this direction are the depots of the Southern, Western and Northern Railroads—here, also, are many of our hotels. On this side are the Greenwich, North River and Merchants' Exchange Banks, and the Ocean, Irving, Grocers', and Knickerbocker Banks, more recently established. Several streets in this section have been lately improved and widened, augmenting the value of real estate, while many old buildings have been superseded by stores not surpassed in elegance or capacity by any in the city. The docks and wharves must multiply with the continuing growth of the port, (A) bringing the other side of the river into greater use. Weehawken Ferry, while enhancing the value of the contiguous property, will relieve the pressure from the commercial centre, which pressure must continue to extend to the adjacent shores.

Ferry boats are of great service between the metropolis and its outskirts. They are convenient and safe. Experience proves that they are invaluable as mediums of communication; and confessedly the quickest and most pleasant method of going the same distance. No bridge or tunnel could equal them for convenience. As New York has progressed, ferries have multiplied and improved. Their number is already considerable, and an accumulating supply must continue to meet the exigencies of the increasing demand. (B)

The route of the plank road is through some of the richest country near New York, suitable for the cultivation of garden truck in an eminent degree, and affording very many beautiful seats for country residences. It will also be an outlet for the produce of a hitherto almost unknown region, rich in agricultural products and mineral wealth. With an easy grade, and the abundance of timber in the vicinity of the Ramapo, it can be constructed at a moderate expense; the whole of the cattle droves, which now land at the lower part of the island, and then are driven through the streets to the upper part of the city, can be landed directly at the highest point practicable, (say 40th street;) and gardeners and farmers can also the saved much expense and hard travel by taking the ferry at Weehawken, rather than those at Hoboken and Jersey City.

The City of New-York is assuming a most conspicuous position. In population and commercial importance, it is increasing with unparalleled speed. It has become the first city on the continent, and the third in the world. Possessing a safe harbor of unequalled extent, contiguous to the ocean, with navigable rivers running from many directions to it, (C) the city has unsurpassed advantages conducing to its progress. Connected by lines of railroads with extensive agricultural regions, and having telegraphic communications with all important points of the Union, it may be termed the heart of the western hemisphere. Contemplating its past progress, (D) we find, when ceded to the English in 1674, the city contained a population of about three thousand. Nearly three quarters of a century later, in the year 1748, the city numbered twelve thousand inhabitants. In 1786, the city numbered upwards of twenty-three thousand inhabitants.

New-York City, under the genial influences of a republican government, became still more prosperous and thrifty. At the commencement of the current century, the population of New-York City was above sixty thousand; that of the country at large being five and a half millions. Twenty years passing, with an intervening war, in 1820, the population was nearly one hundred and twenty-four thousand; that of the Republic being about nine and a half millions. In the year 1830, the population of New-York City exceeded 200,000; and in 1840, ten years later, it exceeded 300,000. Its commerce had increased in activity, and its wharves were crowded with ships of all classes. (E) Suburban cities and villages, increasing with celerity in extent and population, participated in the prosperity of the port. Brooklyn, in 1830, numbered 15,394; in 1835, 24,529; and in 1840, 36,233. Williamsburg, Jersey City and Hoboken, rapidly augmented their numbers; and dwellings and lands not a quarter of a century previous surrounded by the quietness of rural life, yielded to the encroachments of the metropolis. In 1845, the population of New-York City was 371,223 souls; being an increase of about 70,000 in five years, or 14,000 per annum. In 1850, the latest official census, the population was 515,394; being an increase of about 144,000 in five years, or nearly 30,000 annually. The environs also exhibited prolific increases in proportion. (F) Taking the aggregate population of the city and suburbs as indicated by the late census, we had, in 1850, a population of about seven hundred thousand; this is rapidly, almost incalculably increasing. The number and magnificence of the hundreds of new buildings which are annually erected, show animated progress. (G) The harbor is well defended, and commercial relations are rapidly extending. (H) Immigration is continually increasing; (I) the arts and sciences, and every kind of learning are in the ascendant. The streets and ways of the city are crowded; and multitudes of vehicles bear to and fro its commodities and denizens. Communication is increasingly demanded with contiguous regions; and active enterprises are needed to keep pace with an expanding population. Inter-communications with the environs must improve and be constant. The growth of the whole country being promoted by the net work of rail-ways and canals spread over it, is evidence that a similar principle applies to the city and its suburbs. Easy and speedy ingress and egress are especially important. Means for these purposes must be multiplied without stint in the ratio of an increasing population. The practical results of the elementary principle, "time is money," must be continually borne in mind; and our inhabitants must have the opportunity to reach every direction with speed and facility.

"New York is the London of the western world, where all great operations of exchange concentrate, receiving constant tribute from the industry of the world. Here the cotton, the flour, the rice, the corn, the tobacco, the manufactures, and the gold of the whole country meet the products of all other lands and climes, and here the multifarious transfers are made. Here is concentrated the capital that puts in motion the power of steam and the strength of muscle in the remotest districts—that stretches the iron track and impels the rapid engine to the extreme verge of civilization. When we compare the present expansive power and wealth of the Union, and the extent of commercial connections of this emporium with the picture they present to the memory of many now living, it seems perfect soberness to anticipate that within the periods of lives now in being, the centre of power will have been assumed by our government, and commercial supremacy universally conceded to New York."*

The rapid growth of the City of New York would seem, at this time, especially to demand a locality such as Weehawken, where the population can find places for both cheap and elegant residences accessible, surrounded by a pure and healthy atmosphere, and panoramic views of the city and surrounding country of almost unequalled beauty; and which, while possessing all these advantages, is likely to increase rapidly in value from the transportation of passengers and freight upon the road and ferry, and the growth consequent thereupon.

The prospects of New York are cheering and encouraging. The boldest flights of imagination, fifty years ago, could scarcely have pictured the realities of the present hour. "People still live," says a late paper, "who can recollect the Empire City when it was a mere cluster of houses around Nassau and Pearl streets. Many a man, yet in the full vigor of life, has shot snipe in the marsh through which Canal street runs. Old ladies tell of the splendid balls which used to be given in their youth when John street and William street were the centres of fashion. But a few months ago, Union Square was the extreme limit of civili-

* *N. Y. Times*, March 27, 1852.

zation. You will now find that your acquaintances are rapidly moving above Fourteenth street. Elegant and substantial houses are going up in Fifty-first street; and lots in 121st street have been sold for $1,100." Observing the past and present, it may be safely said that in much less than 20 years New York City will have doubled its present population and trebled its wealth. Many reasons conspire to favor the impression that the proportional increase will be greater in future than during the past. The present prosperous condition of our country, the dissatisfaction among the masses of the old world, the uncertain condition of European affairs, all urge to this conclusion. Upwards of 300,000 immigrants arrived at the port of New York last year. Many of these, it is true, take up their abodes in rural regions and distant parts of the country, yet many also locate themselves at this focus of trade. The accession of population from the country is also very great. Numbers of young men from the surrounding rural districts come here to seek their fortune. Here, then, the question arises, where are the overflowings of this increasing mart to go? The answer is plain: they must overspread the soil nearest to the commercial heart of the metropolis. They will take up their abodes as close as practicable to the localities of traffic, and as accessible as can be to the streets of trade. To enable this to be done, and to facilitate ingress and egress, resort will be had to every expedient that will afford communication with all points of the surrounding region; our citizens must proceed from business to their homes with facility; and ferry boats will ply to and fro, to an extent not now thought of.

Bearing in mind, the natural advantages of Weehawken, and the facts that the business of New-York is centering on the Hudson, and that the city can only extend in a northerly direction, we hazard nothing in predicting the speedy growth of a large city upon the beautiful and extensive country which lies in the neighborhood of the Weehawken ferry.

NOTES.

(A) From January 1, 1851, to January 1, 1852, the number of vessels entering this port was 3,888, of which 2,381 were American. The number of passengers brought in these vessels from foreign ports, was 299,081. The daily arrivals at the port of New York are nearly double those at any other port in the Union. *N. Y. Herald, April 7th*, 1852. A city almost as large as Philadelphia is annually emptied from ships upon the New York docks. The increase of American population, by immigration, is now half as great as its natural increase. And every thing indicates this ratio will continue to advance. *N. Y. Times, April 2nd*, 1852. Canal navigation, Railroads and Telegraphs bring New York into direct communication with every part of the land; and show that her commercial interests are in a rapid condition of progress. *See Belden's N. Y., p.* 140. Over 600,000 passengers went over the Hudson River Railroad during the first nine months it was in operation. When the road shall be completed to Albany, the travel will be still further and largely augmented. *Hunts' Mer. Mag., Jan.*, 1851, *p.* 125. New York is the grand focus whence the system of electric telegraphs radiates to all parts of the Union. She is the point where the Union centers. The advantages of this great engine of intelligence are, therefore, pre-eminently hers. *Idem, Aug.*, 1852, *p.* 168. The "West India Docks," situated at Blackwall, on the river Thames, are about 3½ miles from the London Exchange, or the centre of the main business of London. *Atlantic Dock Prospectus.*

The growth of New York will be either retarded or assisted, just in proportion as the inter-communication with its suburbs is obstructed and contracted, or enlarged and made easy to the inhabitants. *N. Y. Herald, March* 26, 1852.

(B) There are 18 ferries between this city and its environs. 1 to Bulls Ferry and Fort Lee, from Canal street; 3 to Hoboken, from Christopher, (1 mile) Canal, (1½ miles) and Barclays streets, (2 miles); 1 to Jersey City, from Courtlandt street, (1 mile); 1 to New Brighton and Port Richmond, from Battery Place; 1 other to Staten Island from Whitehall; 1 to Greenwood, from East side of the Battery; 5 to Brooklyn—1 from Whitehall to Hamilton avenue, 1 from Whitehall to Atlantic street, (1,472 yds.,) 1 from Fulton street, 1 from Catharine street, (736 yds.,) and 1 from Governeur street, (707 yds.,) 4 to Williamsburgh—1 from Peck slip, (2800 yds..) 2 from Grand street, and 1 from Houston street; and 1 to Astoria from foot of Eighty-sixth street. *Valentine's Manual*, 1852, *p.* 220.

(C) The Harbor of New York is spacious. Upon its bosom, surrounded by protecting headlands, the combined navies of the world might float in perfect safety. The depth of water at the wharves is sufficient for the largest vessels. The mild temperature and the strong currents that sweep through the adjacent rivers, complete the natural advantages of the harbor. For a century it has been but twice blockaded with ice. *New York: Past, Present, etc., p.* 60. The Bay of New York communicates with Newark Bay through the Kills, in the west, between Staten Island and Bergen Neck, and with another bay at the south, called the Outer or Lower Harbor, through the Narrows, a compressed strait, between Staten and Long Islands. This latter bay opens directly into the ocean. The inner harbor, as well as being one of the best, is also one of the most beautiful in the world. *Hunts' Merchants' Mag., Aug.,* 1852. The Hudson river affords an inland navigation, in the Empire State, one hundred and twenty miles longer than the Thames. The harbor of New York receives the vessels that navigate rivers, canals and lakes, to the extent of three thousand miles, equal to the distance from America to Europe.

(D) In 1673, almost all the houses in New York presented their gable ends to the streets. All the most important or public buildings were set upon the foreground, to be more readily seen from the river. The chief part of the town of that day lay along the East river, (called *Salt* river in early times) and descending from the high ridge of ground along the line of the Broadway. (*Watson's Sketches.*) In 1677, there were 12 streets and 384 houses in the city. *Hunts' Merchants' Mag., Aug.,* 1852, *p.* 165.

(E) Official statements exhibit, in 1840, the English tonnage to be but trifling over the American. It is not an uncommon occurrence in New York for merchant vessels to be detained in port from one to fifteen days before they can secure a suitable berth for discharging; and then, often times, the best they can obtain is an outside one, obliging them to discharge their cargoes over the decks of two or three other vessels. *Prospectus of Atlantic Dock Co.,* 1840.

(F) Brooklyn, in 1845, numbered 59,574; in 1850, 97,838.—Williamsburg, in 1840, 5,094; in 1845, 11,338; in 1850, 30,780. Jersey City, in 1840, 3,033; in 1850, 6,856. Hoboken, in 1850, 2,678. Bergen Township, in 1850, 2,801; North Bergen, in 1850, 3,576. Newark, in 1840, 17,290; and in 1850, 38,894.

(G) Buildings erected in N. Y. City, in 1846, 1,910. In 1847, 1,846; in 1848, 1,191; in 1849, 1,495; and in 1850, 1,912. Aggregate number for ten years (1840-50) 15,409. *Hunts' Mer. Mag.; Vol. XXV., p.* 136. For description and cost of dwellings lately erected, see *Valentine's Manual,* 1851. *Article: "Progress of the City."* Number of buildings put up in Brooklyn, excluding 8th and 9th Wards, 1,371. *N. Y. Herald, April 15th,* '52. Buildings have also rapidly increased in number at Williamsburg, Jersey City, Hoboken and all other regions around New York.

(H) *See Hunts' Mer. Mag., Aug.,* 1852, *pp.* 161-2. During four years following 1624, the exports of N. Y. City were valued at $68,000; the imports at about two thirds of this amount. In 1839, value of exports exceeded $36,000,000; the imports, upwards of $97,000,000; duties collected exceeded $13,000,000. In 1848, value of exports nearly $46,000,000; imports beyond $89,000,000; duties exceeded $19,000,000. *Belden's New York, etc.* Vessels arrived at New York in 1821, 912. In 1830, 1,489; tons, 314,715. In 1840, vessels, 1,953; tons, 527,594. In 1850, vessels, 3,341; tons, 1,247,860. *Hunts' Mer. Mag.: Vol. XXIV, p.* 746. In 1821, the number of arrivals at this port was 912. In 1851, it was 3,840. In 1821, the amount of tonnage was 171,000. In 1851, it was 1,624,000, or almost a thousand per cent increase! Let us suppose that we go on at the same ratio for another thirty years, and we are overwhelmed with the immense calculation and the scene before us. *N. Y. Chr. Intelligencer, April* 8, 1852.

(I) Immigration into the port of New York, in 1849, 234,271; in 1850, 230,620. Taking past per cent. the population will double in about 18 years; in 1863 it will be 742,446; in 1881, 1,484,892. *See Belden's N. Y., p.* 140. The predictions of Mr. Pintard, made in 1806, were deemed wild and visionary. But experience has shown that they were conceived upon calculations worthy of reliance, and that far-penetrating anticipations, judiciously attained, are not unfrequently realized. He predicted a population for New York, in 1840, of 361,195; and in 1850, 564,520. His prediction for 1855, is a population of 705,650; and for 1860, 880,062; for 1865, 1,010,577; and for 1870, 1,378,225.

www.ingramcontent.com/pod-product-compliance
Lightning Source LLC
LaVergne TN
LVHW020635110826
845149LV00004B/1196

* 9 7 8 1 4 1 8 1 9 1 5 2 8 *